Economic Prosperity
and
Space Development

Economic Prosperity and Space Development

Michael G. Baran

Library of Congress Control Number:		2010909183
ISBN:	Hardcover	978-1-4535-2617-0
	Softcover	978-1-4535-2616-3
	Ebook	978-1-4535-2618-7

This book was printed in the United States of America.

To order additional copies of this book, contact:
Xlibris Corporation
1-888-795-4274
www.Xlibris.com
Orders@Xlibris.com
82409

as the ice age, of enormous proportions. Not only that, but Earth has also experienced many asteroid collisions, some minor, some catastrophic. We now face the challenge of where we are in the timeline for the next significant occurrence. In addition, who knows? Just as the indigenous peoples of the Americas experienced invasion and colonization by Europeans, do we humans face the possibility of experiencing extraterrestrials visiting or invading Earth? Ultimately, we humans need to be prepared, as much as we can, for what could happen. Having human habitation on another planet in our solar system, even perhaps in another solar system, and developing our technological capabilities further would give us options that don't exist today.

Current Economic Conditions

Market Conditions

I'll let the stock index charts speak for themselves with regard to major stock values for each economic area. Major sectors of the economy, banks, automotive, and various other industries received a major shock with the sudden reversal of what then appeared to be an overstimulated economy. There seemed to be significant overproduction in the housing industry, automotive industry, and high-tech industry for the market at the time. For many of these stock market indexes, they are seeing their values only approaching their 2006 values. But this is not the only reason for the current turmoil being experienced in the world economy. (Note: These charts are courtesy of Yahoo! Finance, Yahoo! Web site.)

The Dow Jones, a Major North American Stock Index

Prime indicators of the economic condition in North America that exist are some main North American trade numbers. "Trade using surface transportation between the United States and its North American Free Trade Agreement (NAFTA) partners Canada and Mexico decreased by 23.3 percent in 2009 compared to 2008, dropping to $637 billion, according to the Bureau of Transportation Statistics (BTS) of the U.S. Department of Transportation. The 23.3 percent decline in trade was the largest year-to-year decline for the 15 years covered by these data."[2]

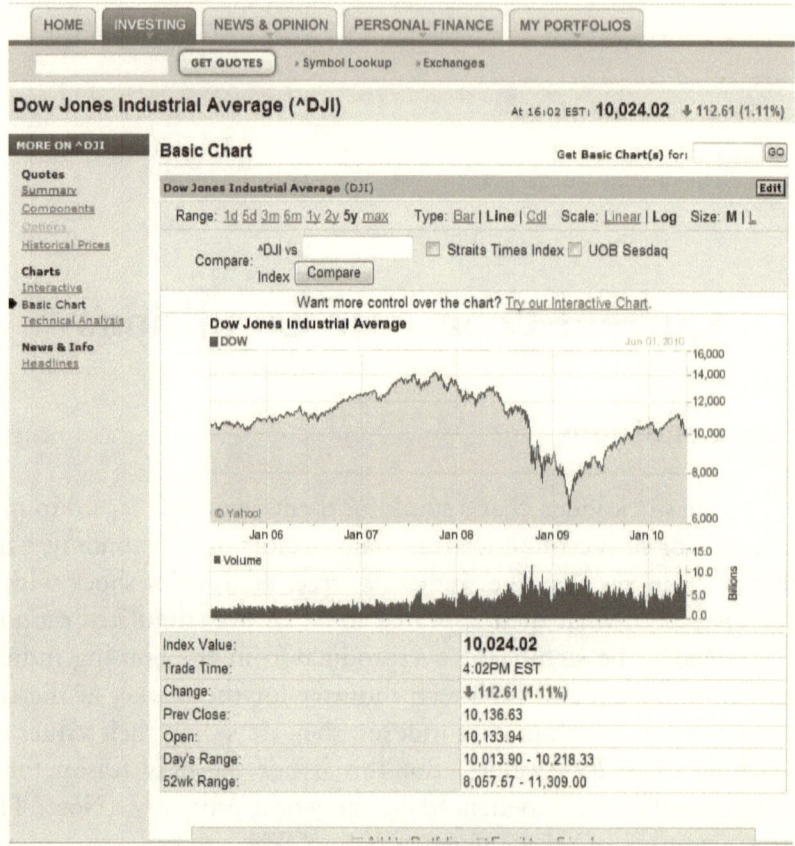

²A1. Dow Jones Index

The value of U.S. Surface Transportation Trade (with Canada and Mexico) totaled $829 billion in 2008. But, also averaged $750 billion from 2005 to 2007. However, due to the economic downturn the Surface Transportation Trade fell to a considerably lower total of $637 billion in 2009. I doubt that the Surface Transportation Trade has gone much above the 2009 number in 2010. "U.S.-Canada surface transportation trade totalled $386 billion in 2009, a decrease of 28.1 percent compared to 2008. The value of imports carried by truck was 25.7 percent lower in 2009 than

---

[2]   http://www.bts.gov/press_releases/2010/bts014_10/html/bts014_10.html   2009   Surface Trade with Canada and Mexico Fell 23.3 Percent from 2008 RITA Research and Innovative Technology Administration Bureau of Transportation Statistics BTS from the US Department of Transportation : March 18 2010

2008 while the value of exports carried by truck was 20.2 percent lower." "U.S.-Mexico surface transportation trade totalled $251.0 billion in 2009, a decrease of 14.4 percent compared to 2008. The value of imports carried by truck was 12.2 percent lower in 2009 than in 2008 while the value of exports carried by truck was 10.8 percent lower."[2]

Bureau of Transportation Statistics BTS
from the US Department of Transportation : March 18 2010

The FTSE, the Index Used to Track Major UK Stocks

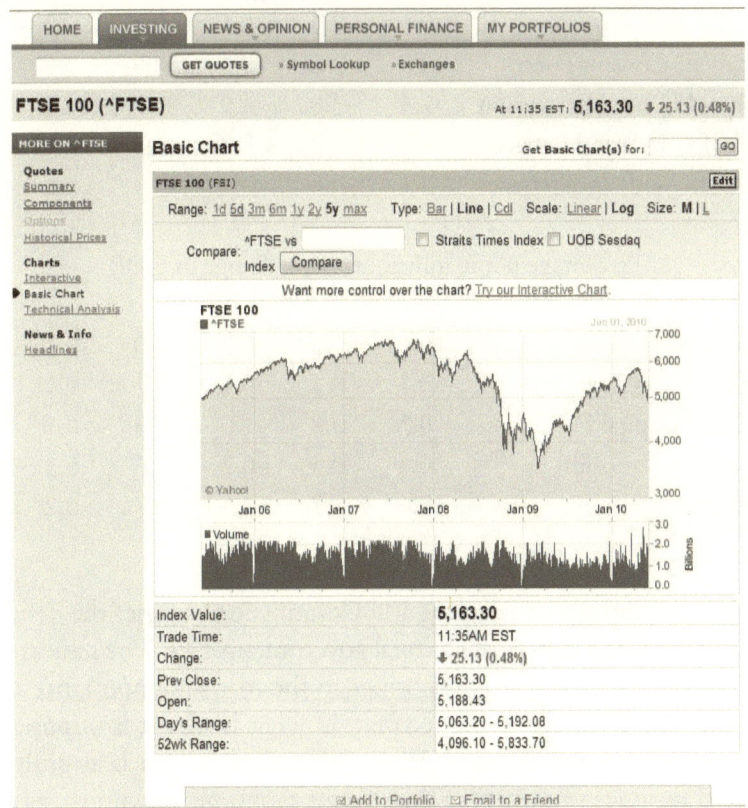

A2. FTSE UK Index

From the European Commission economic reporting, we can see the effect of the downturn and the fragile if temporary recovery on the way. Some of the reporting also reflects world data. The following report indicates the recent negative growth trend from the various economic contributors of the EU economy.

Quarterly Report on the Euro Area[3]   IV/2009

Table I.1.1 Euro-area Growth Components

Description	2008 Q3	2008 Q4	2009 Q1	2009 Q2	2009 Q3	Carryover to 2009
Percentage change on previous period, volumes						
GDP	-0.4	-1.9	-2.4	-0.2	0.4	-4.0
Private Consumption	0.0	-0.5	-0.5	0.0	-0.2	-1.0
Government Consumption	0.5	0.6	0.6	0.6	0.5	2.3
Gross Fixed Capital Formation	-1.4	-3.8	-4.9	-1.7	-0.4	-9.9
Exports of Goods and Services	-1.3	-7.2	-8.7	-1.3	2.9	-14.0
Imports of Goods and Services	-0.1	-4.8	-7.4	-2.9	2.6	-11.8
Percentage point contribution to change in GDP						
Private Consumption	0.0	-0.3	-0.3	0.0	-0.1	-0.6
Government Consumption	0.1	0.1	0.1	0.1	0.1	0.5
Gross Fixed Capital Formation	-0.3	-0.8	-1.0	-0.3	-0.1	-2.2
Changes in Inventories	0.4	0.3	-0.7	-0.6	0.3	-0.6
Net Exports	-0.6	-1.2	-0.6	0.7	0.2	-1.2

The following report from the EU Commission outlines the deficit and debt constraints the collective national governments face. The serious nature of the projected 2010 debt levels brings to the fore the importance of how at risk the long-term economic recovery is. That is why it is important the G20 economic community be open to new strategies (such as outlined in this book) in order to overcome the current economic downturn, as well as develop a strategy for prolonged future economic prosperity and growth.

---

[3]   From the European Commission site http://ec.europa.eu/index_en.htm Source Commission Services   http://ec.europa.eu/economy_finance/publications/qr_euro_area/index_en.htm (Quarterly report on the euro area. 4. December 2009. Brussels. PDF. 66pp. Tab. Graph. Free.) ISSN: 1830-6403

As a consequence of the steep fall in revenues, fiscal stimulus measures under the European Economic Recovery Plan (EERP) and the operation of automatic stabilisers, government balances have deteriorated sharply. Thanks to effective policy action since autumn 2008, coordinated in the context of the EERP, a financial meltdown and a generalised loss of confidence has been avoided. Fiscal policymaking has been successfully targeted to the need and urgency of pulling the economy out of the recession. Discretionary fiscal stimulus and unfettered automatic stabilisers have provided a cushion to economic activity and contributed to the recent signs of improvement, but have led to a substantial deterioration in government accounts. Rising budget deficits and low or negative growth, as well as the support being given to the banking sector, are feeding into significantly higher public debt levels. The average euro area budget deficit is now expected to increase from 2% of GDP in 2008 to over 5% of GDP in 2009. On the basis of current plans and projections, the euro area deficit will further increase to 6½% of GDP in 2010, while public debt could reach 84% of GDP by 2010, i.e. an increase of 18 percentage points from 2007. In 2009, with the possible exceptions of Cyprus and Luxembourg, almost all euro area Member States will post budget deficit ratios above the 3%-of-GDP threshold, with some countries exceeding the benchmark by a large margin. In the first half of 2009 and on the initiative of the Commission, the Council opened Excessive Deficit Procedures (EDPs) for Greece, Ireland, France, Malta and Spain on the basis of a breach of the reference value in 2008 (2007 for EL)1. The Commission proposes today that the Council open EDPs for countries which are expected to breach the reference value in 2009. The flexible application of the Excessive Deficit Procedure permitted under the Pact provides important support and direction for Member States in these difficult circumstances. As a result, budgetary consolidation paths recommended under the EDPs have been largely set in a medium-term perspective and, depending on circumstances of individual countries, longer deadlines have been recommended for correction of the excessive deficits.[4]

---

[4]   http://ec.europa.eu/economy_finance/publications/publication15951_en.pdf   European Commission Annual Statement on the Euro Area - 2009 Economic and Financial Affairs Directorate (pg 3-4 in report)

The impact of the financial crisis fed almost instantly into the real economy. During the fourth quarter of 2008, survey indicators of global economic activity dropped steeply. The OECD leading indicator fell to its lowest level since the mid-1970s. This was mirrored by a precipitous decline in world trade, which dropped by some 6% in the last quarter of 2008, a rate not registered since World War II (Graph 1.1). Industrial production was severely affected as well, with double-digit contractions both in the main industrialized regions and in many emerging market economies.[5]

From a global perspective with real procurement data, we can see the drastic drop in reflective industrial production in the following graph with a tentative recovery appearing in 2009 in certain areas of the world's economy.

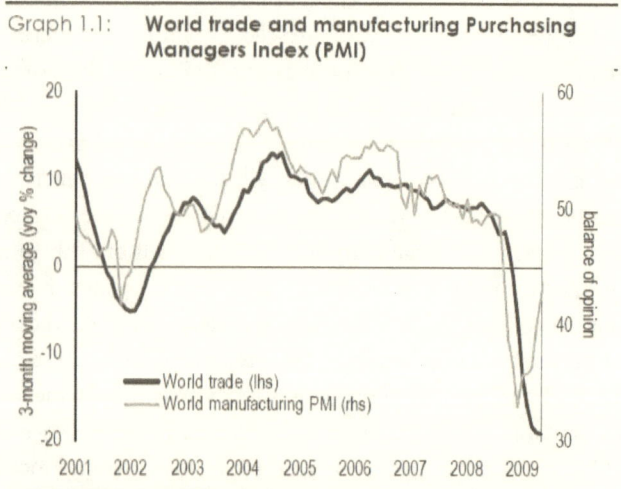

Graph 1.1:   **World trade and manufacturing Purchasing Managers Index (PMI)**

Note: Data for world manufacturing PMI cover the period January 2001 to May 2009
Source: CPB Netherlands Bureau of Economic Policy Analysis, EcoWin, Bloomberg.

A3. World Trade and PMI[5]

[5]   http://ec.europa.eu/economy_finance/publications/publication15951_en.pdf
European Commission Annual Statement on the Euro Area—2009
Economic and Financial Affairs Directorate (Page 17 1.4 of document) (pg 25 of .pdf)
MACROECONOMICDEVELOPMENTSECONOMICSITUATIONANDPROSPECTS

The Nikkei Index Charting the Major Japan Stocks

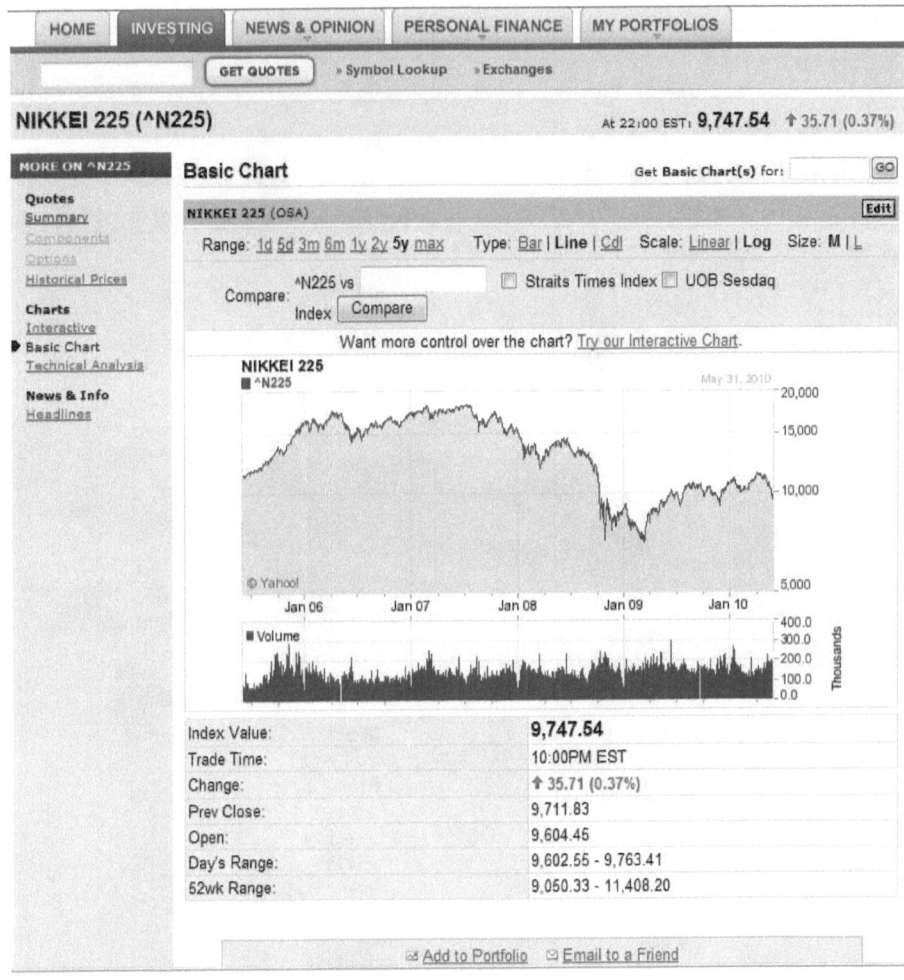

A4. Nikkei Index

The Hang Seng, the Hong Kong Stock Index, Major Chinese Stocks

A5. Hang Seng Index

The Mumbai Index (Bombay), the Major India Index

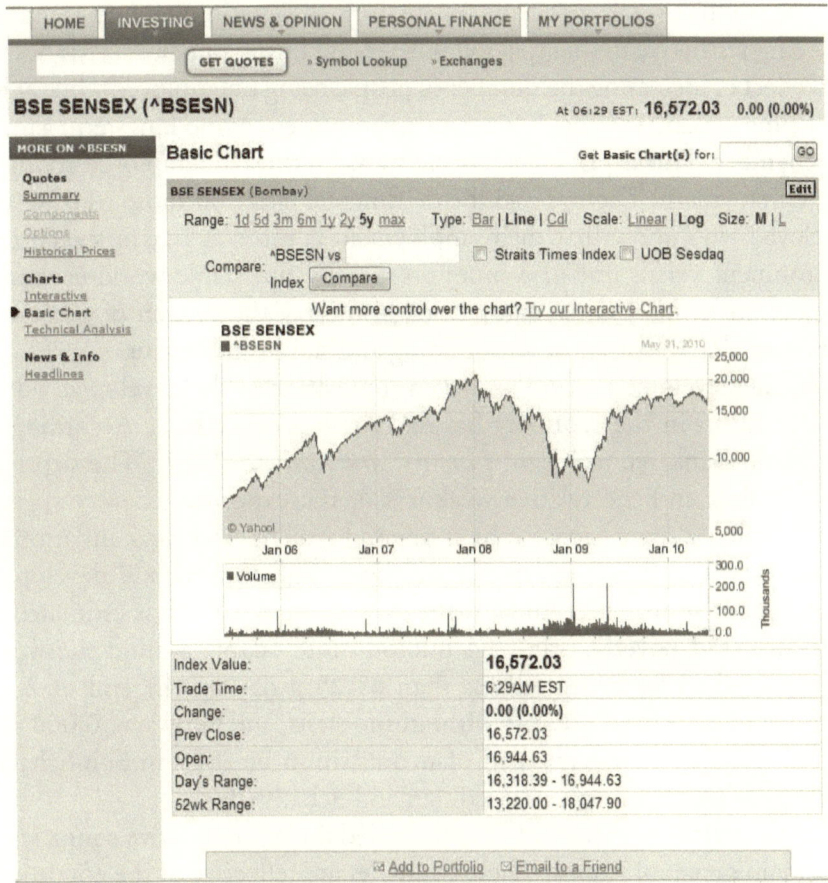

A6. BSE Bombay Index

These indexes in part provide a picture of the current world economic conditions. The lower stock values reflected in the indexes is a result of the decreases in stock values due to poor corporate earnings expectations. We are now facing constrained conditions in employment, city infrastructure financing, and corporate revenues. Many corporations are seeing current and forecasted revenues at risk due to the conditions of the markets drying up in the housing, automotive, and high-tech sectors, perhaps due to an inevitable market saturation in those sectors or perhaps in part due to overly liberal lending policies in the U.S. housing market by some financing parties, or a combination of both.

Employment Conditions

Due to the overall economic uncertainty, corporations and governments are being restrictive in their payroll outlays and hiring practices. This has the dual effect of increasing the number of people being unemployed while at the same time making it difficult for those unemployed to find long-term gainful employment. This is placing a terrible stress on those unemployed trying to find employment. There is also the detrimental effect on those who are still employed who know that their employment is more at risk in the current environment versus one of a more prosperous and stable world economy. The current employed, in many cases, also carry the burden of additional workloads due to corporate streamlining and staff restructuring.

In the developing world and even some areas of the developed world, the downturn in the economy has had some severe effects. According to the World Bank report *Crisis, Finance, and Growth 2010*, "The depth of the recession and the relative weakness of the expected recovery suggest that significant spare capacity, high unemployment, and weak inflationary pressures will continue to characterize both high-income and developing countries for some time. Already, the slowdown in growth is estimated to have increased poverty. Some 64 million more people around the world are expected to be living on less than $1.25 a day by the end of 2010 than would have been the case without the crisis, and between 30,000 and 50,000 children may have died of malnutrition in 2009 in Sub-Saharan Africa because of the crisis" (Friedman and Schady 2009).[6]

In the following reports, I have indicated the representative countries in each main geographic area. These countries are reflective of the conditions for each geographic area—some are better off than their neighbors, some are worse off.

USA National Employment

With the downturn in the housing and automotive sectors of the U.S. economy. The U.S. has lost many long term jobs in these sectors as well as in industries supporting those particular sectors. This is reflected in the job loss data and information provided below.

---

[6]   http://siteresources.worldbank.org/INTGEP2010/Resources/GEP2010-Full-Report.pdf
     Crisis, Finance, and Growth2010 Global Economic Prospects Pg3 Overview

There was a net loss of 216,000 jobs in the month (August 2009), according to the Labor Department, that was the fewest jobs lost since August 2008 and lower than a revised loss of 276,000 jobs in July. Economists surveyed by Briefing.com predicted a loss of 230,000 jobs in August. But even with the lower level of losses in August, 6.9 million jobs have been cut from payrolls since the start of 2008.[7]

"The unemployment rate, which in July fell for the first time in 15 months, turned higher again, jumping to 9.7% from 9.4% in July. This is the highest the unemployment rate has been since June 1983. Economists forecast that the jobless rate hit 9.5% in August." The "report also showed that there were 9.1 million workers limited to part-time jobs because they couldn't find full-time positions, up 278,000 from the month before."[7]

Job losses are expected to continue at least into the middle of next year, likely driving the unemployment rate above 10 per cent from 9.8 per cent last month. It could take three or four more years for it to fall to normal levels. The longest and deepest downturn since the Great Depression has claimed 7.2 million American jobs since it began in December 2007.

Analysts figure 750,000 more jobs could disappear over the next six months. If you add in people who have stopped looking for work, or who are working part time when they want a full-time job, the unemployment rate is a whopping 17 per cent, according to the Labor Department. Another sign of continuing distress, applications for Social Security retirement benefits are up 23 per cent from last year, a much larger jump than in other recessions. Signing up for Social Security benefits as early as age 62 can be an immediate source of income for laid-off older workers, but it's also a troubling sign of the scarcity of jobs.[8]

---

[7]   http://money.cnn.com/2009/09/04/news/economy/jobs_august/
      By Chris Isidore, CNNMoney.com senior writer
      Last Updated: September 4, 2009: 4:31 PM ET
      Source of Data : US Department of Labor
[8]   http://money.aol.ca/article/the-recession-may-be-over-yet-job-losses-endanger-recovery-create-woes-for-democrats/719495/

These job losses represent losses in many sectors of the U.S. economy. While it appears the rate of job losses appears to be slowing down, the road to recovery is not yet apparent to all concerned. Since a large part of the U.S. economy is consumer-driven, those people who have lost jobs are no longer procuring items in the market to the extent that they did when employed. This creates a double drain on the economy from a debt standpoint to provide social support payments to the unemployed as well as the reduced market demand by consumers not out on the market buying various products.

UK National Employment

There were 31 million workforce jobs in June 2009, down 163,000 on the quarter and down 664,000 over the year. Most sectors showed falls in jobs over the quarter with the largest falls occurring in finance and business services (down 67,000) and construction (down 61,000).[9]

The unemployment rate was 7.9 per cent for the three months to July 2009. It has not been higher since the three months to November 1996 and it is up 0.7 over the previous quarter and up 2.3 over the year. The number of unemployed people increased by 210,000 over the quarter and by 743,000 over the year, to reach 2.47 million.[9]

Claims for unemployment benefit in August grew by 24,400 from July to 1.61m, the highest since May 1997. And one in five people aged between 16 and 24 is now looking for work, the highest on record, the data showed. The number of jobless in this age group rose from 928,000 to 947,000—edging closer to the landmark of one million and adding to fears of a new "lost

---

The recession may be over, yet job losses endanger recovery, create woes for Democrats
Source: The Canadian Press Posted: 10/12/09 4:44AM
Filed Under: Business News
[9]   http://www.statistics.gov.uk/cci/nugget.asp?id=12
http://www.statistics.gov.uk/default.asp
Office for National Statistics (UK)

generation" of young people. "Unemployment is still likely to reach three million in 2010, and could go higher, said economist Howard Archer of IHS Global Insight."[10]

With unemployment among the young at a significant number above average, there is likely to be a percentage of those that aspired to further education. Without gainful employment, further education becomes unaffordable, which puts more of the population at risk in lacking marketable skills.

Japan National Employment

It is estimated that there are up to 6.07 million surplus workers for all industries, slightly less than 10 percent of the nation's employed workforce. The ratio of job openings to job applicants for July fell to an all-time low of 0.42 for the 14th straight monthly decline. July also saw the number of unemployed people jump by 1.03 million from a year earlier to 3.59 million—marking the largest-ever increase. Of the total, 1.21 million workers had been laid off, 650,000 more than a year earlier.[11]

The number of employed persons in July 2009 was 62.70 million, a decrease of 1.36 million or 2.1% from the previous year. The number of unemployed persons in July 2009 was 3.59 million, an increase of 1.03 million or 40.2% from the previous year. The unemployment rate, seasonally adjusted, was 5.7%.[12]

"Declining demand for full-time workers is worrisome. Full-time job offers in manufacturing have halved in the past year, while offers for part-time and seasonal employment have risen."[11] The demand for full-time

[10]   http://news.bbc.co.uk/2/hi/business/8258405.stm

[11]   http://search.japantimes.co.jp/cgi-bin/ed20090908a2.html
        Tuesday, Sept. 8, 2009 EDITORIAL Rising unemployment rate

[12]   http://www.stat.go.jp/english/data/roudou/154.htm
        Japanese Government Ministry of Internal Affairs and Communications Statistics Bureau,
        Director-General for Policy and Planning & Statistical Research and Training Institute Labor
        Force Survey Monthly Results July 2009 Released on 28 August 2009

workers, I'm sure, is a concern in most economies today, with little certainty as to when this trend of full-time job declines will reverse.

With much of the Japanese economy dependent upon exports, Japan's economic and employment recovery will likely be achieved once the world economy rebounds, which may not happen at least for one to two years if the IMF and World Bank's forecasts are accurate.

China National Employment

While there is limited data on the current Chinese employment situation, we can gather from the following information that the employment status in China can be considered stable to at risk.

> Fiscal stimulus is centered on the infrastructure-oriented (RMB 4 trillion) stimulus plan and the monetary stimulus has led to a surge in new bank lending. Government-influenced investment has soared. Market-based investment has lagged, although positive signs have emerged in the real estate sector. Consumption has held up well. Very weak exports have continued to be the main drag on growth, but import volumes have recovered in the second quarter of 2009 as raw material imports rebounded.[13]

> Prospects for real estate activity appear reasonably good, but consumption is unlikely to pick up speed. In all, China's growth is unlikely to rebound to very high single digit rates before the world economy recovers. We project GDP growth of 7.2 percent in 2009 and 7.7 percent in 2010.[14]

> There is a limit to how much and how long China's growth can diverge from global growth, given that China's real economy is

---

13   www.worldbank.org/china
     World Bank Office, Beijing
     Quarterly Update June 2009
     World Bank China Rpt CQU_June2009_full_06-18-09.pdf pg 2 2nd paragraph

14   www.worldbank.org/china
     World Bank Office, Beijing
     Quarterly Update June 2009
     World Bank China Rpt CQU_June2009_full_06-18-09.pdf pg 2 4th paragraph

relatively integrated in the world economy. Government influenced spending only makes up one-third of domestic demand.[14]

(From the World Bank office, Beijing, quarterly update, June 2009, page 14.)

Table 4. Fiscal developments
(percent of GDP)

	2007	2008	2009 1/
Budgetary revenue	19.9	20.4	18.6
Budgetary expenditure	19.3	20.8	23.4
Budgetary balance 2/	0.6	-0.4	-4.9

Source: CEIC, staff estimates.
1/ Illustrative scenario. See text for assumptions.
2/ Excluding social security and extra-budgetary funds.

Depending on the world economy and GDP growth, China may also be at risk with employment since the deficit situation in China appears also to be growing relative to the current economic malaise. The deficit situation is bound to affect any further fiscal economic stimulus.

India National Employment

The Indian economy grew 6.7% in the year to the end of March 2009, but had grown by an average of 8.8% in the previous five years. Agriculture, which makes up about a fifth of the economy, was one of the sectors to see growth fall, while industrial firms such as Tata have been severely affected by the freeze in world credit markets and a general fall in global spending. In the budget, the government also increased spending on urban poor schemes and the jobs-for-work scheme to help the poor.[15]

Like China, India's economy is still growing but is facing fiscal budgetary challenges. India is also providing much-needed fiscal stimulus

---

[15]   http://news.bbc.co.uk/2/hi/in_depth/business/2009/g20/7897719.stm
Page last updated at 16:16 GMT, Thursday, 24 September 2009 17:16 UK
G20: Economic summit snapshot India

to keep the economy moving. India also has challenges in terms of their very large population and, hence, workforce to keep gainfully employed. However, like the United States, India appears to have a relatively strong consumer-based economy. That may be why the Mumbai (Bombay) index appears to be recovering quite well.

Russia National Employment

> Russia's economy will shrink by 7.5% in 2009, President Dmitry Medvedev has said—but claimed Kremlin intervention had prevented a worse decline. Russia, which is heavily reliant on oil exports, has been hit by the sharp fall in energy prices. Mr. Medvedev said the decline was "very serious" and admitted the government had been surprised at how severely Russia had been hit by the crisis. The most recent official figures suggest that in the year to August, Russian GDP declined 10.2% compared with the same period in 2008.[16]

There is the perception that much-needed investors in Russia are reluctant to invest due to the current fiscal and economic uncertainty. This in turn we can imagine is resulting in increase in the shortage of full-time work available to Russians, which also would affect budgetary issues relating to fiscal stimulus plans.

National and International Debt

A publication by the IMF (April 2009), *World Economic Outlook: Crisis and Recovery*, has indicated that "global activity is projected to contract by 1.3 percent in 2009. This represents the deepest post-World War II recession by far . . . . Output per capita is projected to decline in countries representing three quarters of the global economy. Growth is projected to re-emerge in 2010, but at 1.9 percent it would be sluggish relative to past recoveries . . . . Subject to a number of assumptions, credit write downs on U.S.-originated assets by all holders since the start of the crisis will total $2.7 trillion, compared with an estimate of $2.2 trillion in the January 2009 GFSR Update. Including assets originated in other mature market economies, total write-downs could

---

[16]   http://news.bbc.co.uk/2/hi/business/8301333.stm
        Page last updated at 10:40 GMT, Sunday, 11 October 2009 11:40 UK
        Russia economy 'to shrink 7.5%'

reach $4 trillion over the next two years, approximately two-thirds of which may be taken by banks. Overall credit to the private sector in the advanced economies is thus expected to decline during both 2009 and 2010."[17]

In addition, "fiscal deficits are expected to widen sharply in both advanced and emerging economies, on assumptions that automatic stabilizers are allowed to operate and governments in G20 countries implement fiscal stimulus plans amounting to 2 percent of GDP in 2009 and 1½ percent of GDP in 2010. The current outlook is exceptionally uncertain, with risks still weighing on the downside."[18]

What I am proposing, with the use of the World Bank and national financing, space credits funding and a corresponding huge development in space should allow the world economies another avenue to generate tremendous revenue growth. In some industry sectors, it may mean adapting to the requirements of the space industry. In other sectors, it will just tie in to their current strengths. But for many industries concerned, the growth potential in space development is virtually unlimited. The revenue growth in turn will help impact positively the debt situations of various national governments. With a return to more corporate industrial activity and consequent decrease in unemployment, this will alleviate to a large extent the debt burden many nations face. At the very least, it will incrementally reduce the debt burden and provide long-term opportunities for future, most likely large-scale, tax-base revenues.

While the world's economies are currently still essentially strong, they are being underutilized. They are being underutilized in the sense of market saturation especially in the automotive and high-tech sectors. If these conditions persist only for a short few years, the economy is likely to rebound. But without the added stimulus of a large funding commitment to space development, and soon, the world's economies and nations may experience some long-term detrimental results, part of which will be oppressive debt burdens.

Urban and Industrial Conditions

With the current economic uncertainty, we now have major urban centers facing significant financial circumstances, large populations with increasing unemployment, and huge infrastructures to support.

---

[17] http://www.imf.org/external/pubs/ft/weo/2009/01/pdf/text.pdf 24Apr09
World Economic and Financial Surveys World Economic Outlook Crisis and Recovery Apr09 IMF pg 9 Prospects

The cities listed below are indicative of the challenges we now face in the major economic regions of the world. But these details do not show all the challenges we face. Where there are budget numbers indicated, it is a telling point the sheer size of the budgets involved to maintain the cities in each economic community. The main factor to consider is, in the current economic environment, the tax revenues are bound to be at risk while at the same time, expenses are most likely to rise, which in turn will have a negative effect on the funding for infrastructure maintenance as well as overall improvements to cities' design and environmental integrity.

New York, USA

New York City has a population of 8.3 million and for greater NYC of 16.5 million people (2007 est.). New York in its NYC government publications has indicated some serious financial projections.[19]

NYC Revenue and Expenses May 2008 Plan
In Billions $ Note: FY08 and on are forecasts

	FY06	FY07	FY08	FY09	FY10	FY11
Revenue	37.1	42.0	43.3	40.2	41.8	43.9
Expenses	36.9	41.1	43.4	43.4	44.1	48.8
Surplus/(Deficit)	0.2	0.9	(0.1)	(3.2)	(2.3)	(4.9)
Cumulative Surplus/(Deficit)	3.7	4.6	4.5	1.3	(1.0)	(5.9)[20]

In order to keep the budget in balance (GAAP), NYC is planning to cut back on some capital projects and look at other ways to generate tax revenues. Since the property tax base in most instances is a reliable source of revenue, this should not greatly affect the city revenues. However, other factors such as unemployment may affect the expense side. NYC is projecting employment loss of 175,000 in 2009 and 100,000 in 2010 mostly in the financial high-income sector. [21]

---

[19]   http://www.nyc.gov/html/dcp/html/census/popcur.shtml
       NYC census bureau estimates adopted Dec 2008
[20]   http://www.nyc.gov/html/omb/downloads/pdf/sum5_08.pdf
       The City of New York Executive Budget Fiscal Year 2009 Pg 24
[21]   http://www.nyc.gov/html/omb/html/publications/publications.shtml
       City of New York, Michael R Bloomberg, Mayor. January 2009 Financial

London, UK

London, with a population of 3 million and greater London 7.5 million (2007 est.), faces additional challenges.[22] It appears London is forecasting additional spending and borrowing over the next few years. In 2007 TfL (Transport for London) secured a 10-year funding settlement covering the period to 2017-18. This £39.2bn settlement covers the funding of Crossrail, support for Tube line upgrades and the rest of TfL's operations. By 2012 there will be an increase of over 10 per cent in the capacity of London's public transport network, increasing to almost 30 per cent by 2018."[23a] It appears TfL has an average net services expenditure of around 4.5 billion (pounds) per annum. Another large London organization the Metropolitan Police Authority (MPA) is forecasting a slight increase from 2009 3.1 billion (pounds) to 2012 3.2 billion (pounds) in part to inflation and in part to the upcoming 2012 Olympics.[23b] This is in order to add to and revitalize the Transport for London and the Metropolitan Police Authority in support of the regular infrastructure and significantly in preparation for the 2012 Olympics.

Summary of Consolidated Budget (Pounds)[24]

	2009-10	2008-9
Estimated total expenditure	12,242,300,000	11,354,000,000
Estimate of income	8,215,500,000	7,562,500,000
Estimate of reserves to be used	823,000,000	642,900,000
Estimated total income	9,038,500,000	8,205,400,000
Budget requirement	3,203,800,000	3,148,600,000

Plan Fiscal years 2009-2013. Office of Management and Budget, Mark Page, Director January 30, 2009. pg 7 of document ; pg 10 of pdf.

[22] http://www.london.gov.uk/gla/publications/factsandfigures/factsfigures/population.jsp
DMAG-Update-14.pdf (Aug 2008) pg 1

[23a] http://www.london.gov.uk/gla/budget/index.jsp#final
0910bud-draft.pdf Jan 28 2009 Summary of final draft consolidated budget 2009-10 pg 37 Section 6.1

[23b] http://www.london.gov.uk/gla/budget/index.jsp#final
0910bud-draft.pdf Jan 28 2009 Summary of final draft consolidated budget 2009-10 pg 31 section 4.4

[24] http://www.london.gov.uk/gla/budget/index.jsp#final
0910bud-draft.pdf Jan 28 2009 Summary of final draft consolidated budget 2009-10 pg 15
http://www.london.gov.uk/gla/budget/docs/0809budget.pdf
0809budget.pdf The Greater London Authority's Consolidated Budget and Component Budgets for 2008-09 pg 50 of document; pg 55 of pdf

Additionally, on the national level, the UK IFC has indicated that the effects of this recession may not be over as early as 2010 but may last into 2014. "The recession is now expected for example by the Bank of England—to be deeper and longer than the Treasury anticipated in the PBR, which would be expected to depress tax revenues and increase welfare bills."[25] Especially when taken into consideration the effect of more unemployment and the institutional rescue packages that have been put into place.

Mumbai, India

Mumbai has an estimated population of 13 million and for greater Mumbai of 21 million. Mumbai's challenge is building up the infrastructure of a modern city to meet the requirements for its recent upswing in prosperity. A large portion of the workforce has to travel long distances in a yet undeveloped mass transit system.

Municipal Corporation of Greater Mumbai
Consolidated Account Headwise Budget for the Financial Year 2009-10

Rs. in thousands

Actual Budget est. Budget Rev est. Budget est.

	2007-08	2008-09	2008-09	2009-10
Total Revenue Income (A)	101,571,563	124,348,893	125,983,585	139,833,950
Total Revenue Expenditure (B)	102,817,700	124,426,479	131,494,229	139,910,930
Revenue Surplus/ (Deficit) (A-B)	-1,246,137	-77,586	-5,510,644	-76,980
Capital Receipts Budget Estimates	23,985,960	85,593,379	49,650,637	91,756,070
Capital Expenditure Budget Estimates	19,801,196	86,191,008	56,365,009	89,859,705
Capital Surplus/ (Deficit)	4,184,764	-597,629	-6,714,372	1,896,365 [26]

Highlights of the Budgetary Estimates 2009-10: "Owing to various financial measures adopted by the Central Government and by the Reserve Bank of India in particular, however, the developing country like ours

---

[25]   http://news.bbc.co.uk/2/hi/business/7984889.stm
       http://news.bbc.co.uk/2/shared/bsp/hi/pdfs/06_04_09_pubfin_update.pdf
       http://www.ifs.org.uk/

[26]   http://www.mcgm.gov.in
       combined_Main_Budget200910 Detail.pdf
       (A) Sect 5.37 pg 297 (B) and on Sect 5.38 Pg 298

has not fallen prey to this world wide recession. Though the basis of our economy is strong enough, effect of the recession on Indian economy is felt to some extent and the rate of economic growth has fallen down noticeably. In the end, this has an adverse effect on the major sources of the municipal finance. During the last 3/4 months, the rate of octroi revenue collection has fallen down from 21% to 15%." "Simultaneously, during last few years M.C.G.M. has undertaken huge projects to provide infrastructural facilities and considerable expenditure will have to be incurred for completing these projects." "During the last year, the accumulated amount under various special funds has been utilized for the capital works projects and now it is proposed to withdraw the fund of about Rs.2470 crores (24.7 billion Rs) from the accumulated amounts under the various special funds."[27]

Perhaps the projected numbers for 2009-10 might be a little optimistic given the current market conditions, which are expected to last well into 2010. However, India does have a relatively strong consumer market portion of the economy to rely on, which will help deflect the impact of the world economic downturn.

Shanghai, China

The population of long-term residents reached 17.78 million, according to a sample survey of 1 percent of the city population in 2005. As part of a city public transportation upgrade in Shanghai, "The government announced early this year it would spend 2.6 billion yuan (US$380 million) in the next three years to purchase new buses."[28] As well, there appears to have been welfare and employment initiatives.

In addition to those initiatives, the Chinese central government has taken steps by increasing spending on large construction projects to boost the stalling economic growth in China. "Chinese policymakers hopes that the 4 trillion yuan ($585bn; £423bn) package will reduce the country's reliance on exports." "China has been battered by a fall in demand for its goods, with exports falling a record 25.7% in February" (Feb. 2009).[29]

---

[27] http://www.mcgm.gov.in
combined_Main_Budget200910 Detail.pdf
Sect 1.28 3. HIGHLIGHTS OF THE BUDGETARY ESTIMATES 2009-10 Pg 30
[28] http://www.shanghai.gov.cn/shanghai/node17256/node17971/node17974/userobject22ai31312.html
[29] http://news.bbc.co.uk/2/hi/business/7938893.stm Mar 12 2009
China factory output growth slows.

Tokyo, Japan

As of October 1, 2007, the population of Tokyo is estimated to be 12.790 million or about 10 percent of Japan's total population.[30] In the mayor's own words to the populace of Tokyo, "It is extremely difficult to put together a budget amid such uncertainty and despair, but then, formulating a very austere budget that is overly prepared for the future would only step up anxieties."[31]

The Tokyo municipal government attacks the main problems head-on with their 2009 budget. The first directives are budgetary initiatives to create 500,000 jobs since the Tokyo government is involved in many areas where direct job opportunities are possible. The government is also considering tax incentives for the use of more clean energy such as solar energy as well as support of infrastructure projects such as the Haneda Airport and Tokyo's three loop roads completion, in addition, further mid- to long-term infrastructure projects such as continuous two-level crossings for railways and measures dealing with torrential rainfall, plus the further greening of Tokyo planting lawns on public schoolyards and increasing roadside trees.[32]

Tokyo has also this challenge in these trying times. The central government, facing similar economic challenges, in order to meet some national requirements, is tapping into ¥300 billion from Tokyo's tax revenues over the next two years.[33]

Moscow, Russia

Moscow, with a population of about 10 million people, is one of largest city economy in Europe that makes approximately 20 percent of Russian

---

30   http://www.metro.tokyo.jp/ENGLISH/index.htm
      http://www.metro.tokyo.jp/ENGLISH/PROFILE/overview03.htm
      Population of Tokyo pg1 Population summary
31   http://www.metro.tokyo.jp/ENGLISH/GOVERNOR/MESSAGE/
      Message to Tokyo Residents Updated on March 2, 2009
      Budget 2009 Explanation 2nd paragraph
32   http://www.metro.tokyo.jp/ENGLISH/GOVERNOR/MESSAGE/
      Message to Tokyo Residents Updated on March 2, 2009
      Budget 2009 Explanation 3rd and 5th paragraphs
33   http://www.metro.tokyo.jp/ENGLISH/GOVERNOR/MESSAGE/
      Message to Tokyo Residents Updated on March 2, 2009
      Budget 2009 Explanation 11th paragraph

GDP annually. As of 2007, the Moscow economy reached R6.73 trillion ($263 billion or $364 billion PPP adjusted). A significant portion of Russia's profits and development is concentrated in Moscow as many multinational corporations have branches and offices in the city. Primary industries in Moscow include the chemical, metallurgy, food, textile, furniture, energy production, software development, and machinery industries. According to the 2002 census, the population of the city was 10,382,754.[34]

In March 2009, by the articles of Russian business newspaper *Kommersant*—due to the worldwide economic crisis that was started in 2008 and spread globally—many of development projects in Moscow (especially in the Moscow International Business Center) is frozen and many of them that was planned may be cancelled, like the ambitious Russia Tower in Moscow City. Many of the development groups were reported in near-bankrupt stage like Mirax Group or AFI Development.[34]

City Budget for 2009

Revenues   R1387110.06 million
Expenses   R1447031.07 million
Surplus/ (Deficit) -R 59921.01 million[35]

---

[34]   http://en.wikipedia.org/wiki/Moscow section Economy Overview
[35]   http://www.mos.ru/wps/portal/SearchMosEng
       General Information >> Moscow: events, facts, figures >> Moscow: events, facts, figures 2008/2009
       Section ECONOMY, CONSTRUCTION, INDUSTRY

Environmental Conditions

Given the current world economic conditions and the resulting shortage of available funding, it would appear that the environment may receive the short end of the stick when it comes to receiving the required infrastructure funding for cleanup and prevention. Due to the current strained national and metropolitan financial conditions, the main focus will likely end up being on economic recovery, employment, welfare, debt, and infrastructure maintenance.

What is needed is a world economy setup in such a way that it can support all the needs of a dynamic modern society, a modern society that makes minimal waste an important part of all sectors of the economy. Secondly, as part of the new space development industry with funding from the G20/World Bank financing and space credits we should also be able to fund two areas for the environment: (1) environmental cleanup of existing waste and pollution and (2) city and industry restructuring to prevent further environmental pollution hopefully mainly through greater efficiencies in production and design.

Forms of Pollution

There are many forms of pollution that need to be dealt with on a global as well as national level. All or most forms of pollution can have an effect, if not directly on us, then on the plant, animal, or aquatic life we share this planet with. This is the very plant, animal, or aquatic life we depend on in some way for our sustenance, right down to the microbes.

Ocean Pollution

Ocean pollution is caused by two main factors, runoff from land and rivers and the dumping of waste. Runoff in the form of toxins from pesticides, fertilizers, and other chemicals contaminate nearby rivers that flow into the ocean, which can cause extensive loss of marine life in bays and estuaries, leading to the creation of dead zones. There are similarities with river and other fresh water pollution to the oceanic pollution indicated here. There are four main types of dumped waste:

i)   Industrial waste can include acids, alkaline waste, scrap metals, waste from fish processing, flue desulphurization, sludge, and coal ash. Although there are better practices in place than in prior years this is still taking place.

ii) Sewage sludge, if sludge from the treatment of sewage is not contaminated by oils, organic chemicals and metals, it can be recycled as fertilizer for crops. It has been (and may still be happening) cheaper for treatment centers to dump this material into the ocean, particularly if it is chemically contaminated.

iii) Dredging contributes about 80% of all waste dumped into the ocean, adding up to several million tons of material dumped each year. Rivers, canals, and harbors are dredged to remove silt and sand buildup or to establish new waterways. About 20-22% of dredged material is dumped into the ocean. The remainder is dumped into other waters or landfills and some is used for development. About 10% of all dredged material is polluted with heavy metals such as cadmium, mercury, and chromium, hydrocarbons such as heavy oils, nutrients including phosphorous and nitrogen, and organo-chlorines from pesticides.

iv) Radioactive waste usually comes from the nuclear power process, medical use of radioisotopes, research use of radioisotopes and industrial uses. The protocol for disposing of nuclear waste involves special treatment by keeping it in concrete drums so that it doesn't spread when it hits the ocean floor. The majority of nuclear waste (that is recorded) in the ocean comes from six submarine reactors, one nuclear icebreaker reactor, and damaged nuclear fuel in the Kara Sea. The rest of the nuclear material in the ocean is solid nuclear waste in concrete drums. In addition many cities are sending their waste to Developing nations to deal with. Some of this waste, from overflowing city dumps, is finding its way into the ocean.[36]

---

[36]   http://marinebio.org/Oceans/Ocean-Dumping.asp section Ocean Pollution

There have been important findings that have been in the news recently that relate to both the Pacific and Atlantic oceans. Scientific researchers have found large bodies of plastic debris in both the Pacific and the Atlantic with both having similar properties. To clean this up will require a great deal of funding and technological innovation.

> Researchers are warning of a new blight on the ocean: a swirl of confetti-like plastic debris stretching over thousands of square miles (kilometres) in a remote expanse of the Atlantic Ocean. The floating garbage, hard to spot from the surface and spun together by a vortex of currents, was documented by two groups of scientists who trawled the sea between scenic Bermuda and Portugal's mid-Atlantic Azores islands. The studies describe a soup of micro-particles similar to the so-called Great Pacific Garbage Patch, a phenomenon discovered a decade ago between Hawaii and California that researchers say is likely to exist in other places around the globe" "On the voyage from Bermuda to the Azores, they crossed the Sargasso Sea, an area bounded by ocean currents including the Gulf Stream. They took samples every 100 miles (160 kilometres) with one interruption caused by a major storm. Each time they pulled up the trawl, it was full of plastic.

> "The plastic bits, which can be impossible for fish to distinguish from plankton, are dangerous in part because they sponge up potentially harmful chemicals that are also circulating in the ocean," said Jacqueline Savitz, a marine scientist at Oceana, an ocean conservation group based in Washington.[37]

Air Pollution

Air pollution can be in the form of solid particles, liquid droplets, or gases. In addition, they may be natural or man-made. Earth's atmosphere is a complex, dynamic, natural gaseous system that is essential to support life. Man-made air pollution can have a detrimental effect on this precious

---

[37]   http://ca.news.yahoo.com/s/capress/100415/science/science_cb_atlantic_ocean_junk
    A 2nd garbage patch: Plastic soup resembling Pacific blight found floating in North Atlantic
    Thu Apr 15, 6:53 AM By Mike Melia, The Associated Press

system. As for example, the now-banned chlorofluorocarbons (CFCs) have harmed and perhaps still do the protective ozone layer. Some major primary pollutants produced by human activity include the following:

i) sulfur oxides (SOx)—especially sulfur dioxide, a chemical compound with the formula SO2, is produced by volcanoes and in various industrial processes. Since coal and petroleum often contain sulfur compounds, their combustion generates SO2.

ii) Nitrogen oxides (NOx)—especially nitrogen dioxide are emitted from high temperature combustion. It is one of the most prominent air pollutants and can be seen as the brown haze dome above or plume downwind of cities.

iii) Carbon monoxide is a colourless, odourless, non-irritating but very poisonous gas mainly produced by vehicular exhaust.

iv) Volatile organic compounds VOCs are a significant outdoor air pollutant.

v) Toxic metals, such as lead, cadmium and copper produced in industrial production.

vi) Radioactive pollutants produced by nuclear explosions and war explosives.[38]

Land Pollution

There are various causes of land pollution. Industrial pollution is mainly caused by construction debris, petrochemical contamination from transport and fuels, and heavy metals and chemicals.

> Construction and demolition debris has to be sorted and removed. Currently, much of it ends up in landfill sites. Contamination from petrochemical, heavy metals, and chemicals can occur at the site of production, in distribution, in customer use and disposal Mining is another source of land pollution from heavy metals and toxic metals.[39]

Agricultural production also can cause pollution: "The use of fertilisers, herbicides, and pesticides in the form of residue on the land and run-off

---

38 http://en.wikipedia.org/wiki/Air_Pollution

into rivers and the water table. More and more land is gobbled up by the need for landfill sites to hide the waste from our junk culture. The refuse in landfill sites may itself cause further problems as leachates (polluted liquids) ooze out into neighbouring land."[39]

A Solution

As indicated above, these various forms of pollution are now causing problems for the environment on land and in the water (oceans and rivers). It is essential that these be cleaned up as the long-term hazards can only be considered detrimental to human life as well as other life-forms on Earth. The accumulation of waste and pollution up to the present has to be dealt with. In many cases, the pollution is on such a large scale that large-scale funding and commitment for cleanup will be required. Fortunately, a portion of the funding indicated in the following section can help while at the same time, the return to economic prosperity in each geographic region would allow more funding from each national government tax base.

We can look at the progressive work being done by the Waste Management Incorporated in coordination with the various cities it supports. The direction this corporation has already taken is indicative of the drive and direction required and promoted by governments, government agencies, and environmental groups.

We can presume Waste Management Company in North America is probably similar in scope to other waste management companies in Europe doing similar work. "Waste Management Co. is the leading provider of comprehensive waste management and environmental services in North America. As of December 31, 2008, the company served nearly 20 million municipal, commercial, industrial, and residential customers through a network of 367 collection operations, 355 transfer stations, 273 active landfill disposal sites, 16 waste-to-energy plants, 104 recycling plants, and 111 beneficial-use landfill gas projects."[40]

The waste storage and recycling industry is now a large part of any advanced economy. To give an example of the scope of the industry, one

---

[39]   http://www.greenfootsteps.com/industrial-pollution-causes.html
      Section Industrial Pollution Causes Land Pollution

[40]   http://www.wm.com/wm/press/ekits.asp
      Waste Management Corporation 2008 Annual Report
      Page 2 Introduction A Company You can Count On

of the largest players is Waste Management Corporation, which is a $13 billion revenue a year company with approximately 45,000 employees (2008 data).

What is important here is the development of the newer technologies, contributing to less degradation in the environment with such facilities as the waste-to-energy plants, recycling plants, beneficial-use landfill gas projects. Where needed, these can be expanded and the technology used improved upon, which is important to our future if we are to reduce the pollution in our redesigned cities. As an indication of the direction Waste Management has taken (I assume with encouragement and consultation with various levels of government), Waste Management "pledged by 2020 to have 25,000 acres of protected land for wildlife habitat and 100 landfills certified by the Wildlife Habitat Council. One year later," they "are already halfway to that goal, with 49 of our landfills certified and a total of 21,000 acres around our landfills set aside for wildlife habitat."[41]

Of course, the most important impact we humans can make is to change our habits of consumption and waste disposal. This will go a long way in the prevention of massive future waste buildup that is non-recyclable.

---

[41] http://www.wm.com/wm/press/ekits.asp
Waste Management Corporation 2008 Annual Report
A Company You can Count On pg 3

Funding for Space Development

Due to the unstable world economic situation we find ourselves in, we need to find a way to bring about economic prosperity, a prosperity that will be strong, sustaining, and will provide long-term growth. We can set the stage for that long-term prosperity by funding space development with World Bank long-term financing and with space credits. This space credit funding and long-term financing would provide the needed stimulus to the world economy in the form of a truly viable space industry. While there is current funding for space exploration and scientific study, the attitude and funding to date are not enough to create a viable space industry.

By adding the space industry to the world economic activities, we will be providing the needed stimulus to the ailing world economy. Once we get to the point of a truly vibrant world economy with a higher employment level, we will hopefully be preventing erratic economic conditions that have led, in the past, to political turmoil, warfare, poverty, and environmental degradation. In addition, as much as possible, everyone in the world should be entitled to access to gainful employment.

Current World Gold Wealth

In order to have perspective on the wealth contained on Venus, in the asteroid belt, and other parts of our solar system, we need to know how much wealth exists on Earth. For convenience, we will just look at the gold that has been mined to date on Earth. But keep in mind, there are other valuable minerals as well, both on Earth and in our solar system.

In 2001, it was estimated that all the gold ever mined totaled 145,000 tonnes. One tonne of gold equated to a value of US$30.27 million as of

February 14, 2009 ($941.35/troy ounces). The total value of all gold ever mined would be US$4.39 trillion at that price.[42]

Given possible future market fluctuations, for the purpose of the following exercise, we shall conservatively estimate the total value of gold on Earth (mined to date) at US$4 trillion. An additional consideration is there has been a considerable amount of gold mined since 2001. Therefore, the US$4 trillion estimate can be considered truly conservative.

Hidden Wealth

Just as in the past, European nations sought out wealth from the Americas and found and developed the land, gold, resources, cities, and industries. We can now reach beyond Earth to obtain the wealth that awaits us. Such wealth as in gold and other valuable minerals that might likely be found and readily obtained in our solar system on Venus, Mars, the planetary moons, and in the asteroid belt.

We can assume that since Venus is approximately the same size and makeup as Earth, there would most likely be gold and other valuable minerals on Venus almost equivalent to Earth's. If we also consider the moons of the various planets and the asteroid belt, it would be pretty safe to conclude that altogether, these could produce gold in the order of $4 trillion. This off-world wealth would be equivalent to that produced on Earth up to 2001. In addition, resources other than gold would likely be found there, which will be valuable for the requirements to build buildings, industry requirements, and space transportation. Essentially, the space development infrastructure.

This is, of course, conditional upon the possibility of modifying the climate of Venus and terra-forming it into another Earth, which would also give the overpopulated Earth an opportunity for expansion and wealth creation on Venus. Who knows? Just by shielding Venus from the Sun with satellites, we could control the temperature on Venus and thereby modify the climate to Earth-like conditions, following which we could modify the chemical makeup of the atmosphere and surface in order to enable terra-forming to take place. While this idea may seem farfetched to some, in the near future, depending on our technological advancements at that time, it may very well be achievable. Also, depending on certain circumstances, it may be an urgent necessity.

---

[42]   http://en.wikipedia.org/wiki/Official_gold_reserves
Wikipedia's original source http://www.gold.org/
In gold knowledge/frequently asked questions World Gold Council

Solar System Data According to NASA[43]

Venus
Distance from the Sun: 108,208,930 km 108 Mill. km
Equatorial Radius: 6,051.8 km Diam. 12,100 km
Volume: 928,400,000,000 km3
Mass: 4,868,500,000,000,000,000,000,000 kg 4.87 × 10/24 kg

Earth
Distance from the Sun: 149,597,890 km 149 Mill. km
Equatorial Radius: 6,378.14 km Diam. 12,756 km
Volume: 1,083,200,000,000 km3
Mass: 5,973,700,000,000,000,000,000,000 kg 5.97 × 10/24 kg

Moon
Distance from Earth: 384,400 km
Equatorial Radius: 1737.4 km Diam. 3,474 km
Volume: 21,970,000,000 km3
Mass: 73,483,000,000,000,000,000,000 kg 0.073 × 10/24 kg

Mars
Distance from the Sun: 227,936,640 km 227 Mill. km
Equatorial Radius: 3,397 km Diam. 6,794km
Volume: 163,140,000,000 km3
Mass: 641,850,000,000,000,000,000,000 kg 0.642 × 10/24 kg

Jupiter
Distance from the Sun: 778,412,020 km 778 Mill. km
Equatorial Radius: 71,492 km Diam. 142,984 km
Volume: 1,425,500,000,000,000 km3
Mass: 1,898,700,000,000,000,000,000,000,000 kg 1,899 × 10/24 kg

Assumptions and Conclusions (drawn from the solar system data)

---

[43]   http://solarsystem.nasa.gov/planets/profile.cfm?Object=SolarSys
    Section Resources Compare the Planets
    http://solarsystem.nasa.gov/planets/charchart.cfm

While these resources may seem out of reach given the technology we have today, in fifty or one hundred years or more, with concerted development efforts on behalf of the world geographic economies, we will most likely have developed our technologies far beyond what we have today. Probably, we will have at least established some form of presence and production in space.

Space Credits

Now since this hidden wealth is not immediately available or accessible, it may take decades or even centuries to get to that point. We need to find a method to finance the development that will allow us to get to where we can readily obtain the wealth (gold and other valuable minerals) from the bodies in our solar system other than Earth.

This is where the G20 economic communities and the World Bank will need to (a) finance the space development and (b) create a space credit currency. In order to create funding for space development. This funding would achieve two things: (1) create the infrastructure and development required for expanding into space and (2) provide the needed economic stimulus for long-term sustained and prosperous growth. The funding should be provided to each major geographic economic community on an equitable basis so that all economic communities can participate in the economic recovery and be a part of the expansion into space. The added rationale to this is the more people and corporations in the world with an active economy, the more people and corporations there will be buying products and technology.

Now, for option b) create a space credit currency. Most likely, this currency would be based upon the estimated value of our solar system's gold (excluding Earth), plus the future consideration of four other solar systems. If Venus turns out not to be a possibility, then five other solar systems (excluding our solar system) could be selected. We would, therefore, be considering in total five solar systems times $4 trillion or a grand total of $20 trillion. The World Bank space credit funding from this value would be a percentage of those five solar systems worth (US$20 trillion), let's say 20 percent of the total, giving a resulting value of $4 trillion (space credits). Effectively, the world economy can be funded with $4 trillion of space credits toward space development. Then when the actual gold and other valuable minerals are actually mined, 20 percent would be paid back to the World Bank, and the 80 percent would go to the nation or the company (on the nation's behalf) doing the mining. So the funding available to be provided over a period of time to Earth's geographic economic communities would be

4 trillion U.S. dollars as space credits. The actual calculation, of course, would have to be determined by the World Bank and the G20 financial community. The added benefit of Space Credits is there would be no time limit, at least until the gold is mined. Nor would there be any interest charges.

World Bank G20 Financing

Like any other national support that the World Bank has previously provided, this funding should not be given without any serious plan of implementation at the national and the economic community level. It would be essential that infrastructure and space development (including R&D) be the main focus. But in addition, the funding for environmental cleanup and city and industry restructuring would need to be included to prevent pollution and prevent cities having to support costly outdated infrastructures.

For those that may be skeptical that such a huge and long-term financial arrangement can be accomplished or even be successful, we only have to look at the Marshall Plan that was implemented in Europe after WWII. There were various financial arrangements made with each European nation. These arrangements were long-term and proved to be successful in bringing Europe back to its feet after the devastation of WWII. Fortunately, in this case, we are not dealing with destroyed cities or industries. The cities and industries, in this case, just need updating to the new requirements of the pending space industry.

Marshall Plan—European Funding

The U.S. "Congress passed as the Economic Cooperation Act in the spring of the following year (1948). The act provided more than $5 billion for the first 18 months of what eventually became a four-year program that would cost the American people approximately $13 billion before it ended in 1952. This sum amounted to between 5 and 10 percent of the federal budget over the life of the recovery program, or about 2 percent of the gross national product over the same period".[44]

---

[44]   U.S. Department of State
Bureau of International Information Programs
http://usinfo.state.gov/ Current Site http://www.state.gov/
The Marshall Plan Rebuilding Europe, Blueprint for Recovery by Michael J Hogan page 10

During the Marshall Plan period, Western Europe's aggregate gross national product jumped by more than 32 percent, from $120 billion to $159 billion. Agricultural production climbed 11 percent above the prewar level, and industrial output increased by 40 percent against the same benchmark. Local resources accounted for 80 to 90 percent of capital formation in the major European economies during the first two years of the recovery program.[45]

The main consideration in all the Marshall Plan financing is that it was a concerted effort of most of the European member states as well as the United States of America. It was a well-planned recovery strategy, with the supportive participation of all the parties concerned and with the overall goal being economic recovery for Europe. But it also resulted in a beneficial economic gain for the USA who ended up providing a substantial amount of the raw materials and products needed for the European recovery.

Recent World Bank Projects

We can now look at more recent activity with the loans provided by the World Bank for international projects, projects that were financed in coordination with country-level efforts in developing essential infrastructure components in the individual countries. While the return on investments may not be immediate or readily apparent, the overall contribution to the country-level economy should be considered in broad terms as well as the eventual long-term contribution thereof these will effect. Not only that, but the nearby countries and the geographic economic community the countries are a part of will benefit as well from having a neighbor whose economy and technology is at a developmentally higher level.

These projects are good indicators of the bilateral efforts required to carry off economic improvements. In addition, the requirement for these efforts to be enacted through the many levels of government (government agencies), business, and the public is a ready indicator of the complexity that is involved in many of these initiatives.

---

[45]   U.S. Department of State
Bureau of International Information Programs
http://usinfo.state.gov/   Current Site http://www.state.gov/
The Marshall Plan Rebuilding Europe , Blueprint for Recovery
by Michael J Hogan pg 14

China Railway Funding

The first recent project we will look at is the Second National Railways Project (Zhe-Gan Line), which had a start date for the World Bank of June 2004. The project was initiated by China and the World Bank working in tandem to support China's railways 2003-2007 development plan. The project development objectives were "(a) improve services provided to customers of the Zhe-Gan railway line; and (b) upgrade the quality of track maintenance on heavily used portions of China Railways' network." "The track is to be upgraded on many short sections totalling about half of the track length. The line will also be electrified throughout. Once it is upgraded, the 200 km/h express trains will be operated by electric multiple units (EMUS), while slower passenger and freight trains will be operated with electric locomotives. Component

2) Ministry of Railways (MOR) also wishes to acquire additional highly mechanized, high-capacity equipment for replacing worn rail and switches, renewing ballast, and other maintenance tasks on heavy-traffic lines. Component 3) MOR will hire consultants to analyze, and make recommendations on various issues for improving business processes."[46]

The funding by the World Bank body, the IBRD (International Bank for Reconstruction and Development), was for the amount of $200 million (twenty-year loan). The total project cost was for the amount of $1.75 billion. The financing for the $1.55 billion came from other bank(s). This specific project was completed according to the World Bank project statement in January 2008. This project, it would appear from an economic benefits standpoint, is a costly endeavor that will support the public, businesses, and industries. "The population of south western China was expected to benefit from the increase in passenger services and reduced travel time. The increase in rail freight transport capacity and reduction in transit time of freight trains was expected to benefit the producers of agricultural, mineral, and manufactured goods in south western China by improving access to markets in eastern China." "The original Zhe-Gan line was a 942-km long, double track, diesel-powered, and allowing a maximum speed of 120 km/h.

---

[46] http://web.worldbank.org/external/projects/main?Projectid=P075602&theSitePK=40941&p
iPK=64290415&pagePK=64283627&menuPK=64282134&Type=Overview
World Bank Project No P075602
Second National Railways Project (Zhe-Gan Line)

This project installed new signaling, realigned sharp curves and carried out other civil works needed to permit higher speeds."[47]

From a quantitative standpoint, the project achieved these immediate achievements:

> i)   The transit time per trip for freight trains (minutes) reduced from 1,400 to 931.
>
> ii)   The transit time per trip for passenger trains (minutes) reduced from 600 to 561.
>
> iii) The transit time per trip for express passenger trains (minutes) reduced from 660 to 336.
>
> iv)   Average track downtime (minutes per day) reduced by 50% reduced from 180 to 90.[48]

Turkey Industrial Technology Project

The main intent of the funding for this 1999 US$155 million (forty-year loan) project P009073 was to facilitate Turkey's doing business with the EU on a more level footing. From a government, industry and technological standpoint, this project will further the gains and progress Turkey has already started upon. The approval date for the project was June 1999 and the scope of the project was completed April 2006.

---

[47]   http://www-wds.worldbank.org/external/default/WDSContentServer/
WDSP/IB/2008/08/21/000333037_20080821002337/Rendered/INDEX/
ICR9180P0756020Box334041B01PUBLIC1.txt
Document of The World Bank Report No: ICR0000918
IMPLEMENTATION COMPLETION AND RESULTS REPORT (IBRD-47460)
LOAN IN THE AMOUNT OF US$ 200 MILLION TO THE PEOPLE'S REPUBLIC OF
CHINA SECOND NATIONAL RAILWAYS PROJECT August 15, 2008 Section 1.4 Main
Beneficiaries and 1.5 Original Components : Zhe-Gan Line Improvements

[48]   http://www-wds.worldbank.org/external/default/WDSContentServer/
WDSP/IB/2008/08/21/000333037_20080821002337/Rendered/INDEX/
ICR9180P0756020Box334041B01PUBLIC1.txt
Document of The World Bank Report No: ICR0000918
IMPLEMENTATION COMPLETION AND RESULTS REPORT (IBRD-47460)
LOAN IN THE AMOUNT OF US$ 200 MILLION TO THE PEOPLE'S REPUBLIC OF
CHINA SECOND NATIONAL RAILWAYS PROJECT August 15, 2008 Section F. Results
Framework Analysis (a) PDO Indicator(s)

In January 1996, a custom union agreement was signed between Turkey and EU, allowing most industrial goods to pass freely between the partners. While access to ECU offered Turkey a unique opportunity to accelerate its development through freer and better access to markets, it also made Turkish firms more vulnerable to international competition." "In 1993, Turkish Government approved a comprehensive policy document, Turkish Science and Technology Policy: 1993-2003. This document set four targets: increasing the number of researchers per 10,000 people from 7 to 15; raising the GERD (gross expenditure on R & D) to GDP ratio from 0.3 percent to 1.0 percent; moving up in the rank of scientific publications from 40th to 30th position; and increasing the share of business in total GERD from 18 percent to 30 percent.[49]

Two main accomplishments of this particular project were to further the initial steps taken by Turkey.

1) "Through the reforms at the TPE, Turkey's Industrial Property Rights Regime is now mostly in compliance with WTO and ECU Standards. TPE has modernized its systems and processes and has improved performance in several key areas: patent and trademark applications have increased significantly, average patent processing time has been reduced, and the number of patent and trademark attorneys and IPR personnel in the country has more than doubled.[50]

2) "The TTGV (Technology Development Foundation of Turkey) has become an important agency for R&D financing in Turkey and has supported about 1,400 SMEs with matching TA grants and 260 R&D projects through matching foreign currency loans with a commercialization rate of 90 percent for all R&D projects. The TTGV has attracted about US$150 million of R&D financing,

---

[49]   http://web.worldbank.org/external/projects/main?pagePK=64283627&piPK=73230&theSit ePK=40941&menuPK=228424&Projectid=P009073
The World Bank Report No:ICR0000322 Project ID: P009073
IMPLEMENTATION COMPLETION AND RESULTS REPORT
(IBRD-44950)
ON A LOAN IN THE AMOUNT OF US$155.0 MILLION TO THE REPUBLIC OF TURKEY FOR AN INDUSTRIAL TECHNOLOGY PROJECT
December 19, 2006 Private and Financial Sector Department and Turkey Country Unit Europe and Central Asia Region Public Disclosure page 3 of document ; page 7 of pdf

mostly from the private sector and a majority of its R&D projects are being put into commercial production and synergy between the industry and academic/research community has increased."[50]

Brazil Sustainable and Equitable Growth Program

This program entailed a series of loans directed at specific segments of Brazil. The intent of these loans was to raise Brazil's sustainable economic growth potential and was directed at enabling the economy to increase employment and reduce poverty. The total loan amount was for US$505.05 million and the start date was early 2004, and the funding completion date was December 31, 2004 (whereas some of the initiatives carried through beyond 2004).

Some of the main initiatives looked at were to "cut customs release times by 40 percent, cut container handling costs in ports by 10 percent, lower road transport costs by about 5 percent and increase non-road transportation by 10 percent. Improvements to the business environment will increase public-private partnership, increase cartel prosecutions by the competition authorities, halve the time to register a business in selected cities, and increase the speed of resolution and the recovery value of insolvent enterprises under the new bankruptcy law. Financial sector reforms will reduce bank overheads, increase financial access and reduce credit risk, accelerate the expansion of the insurance industry, and increase access to bank accounts from 95 million to 103 million people by the end of the program."[51]

---

[50] http://web.worldbank.org/external/projects/main?pagePK=64283627&piPK=73230&theSit ePK=40941&menuPK=228424&Projectid=P009073

The World Bank Report No:ICR0000322 Project ID: P009073

IMPLEMENTATION COMPLETION AND RESULTS REPORT

(IBRD-44950)

ON A LOAN IN THE AMOUNT OF US$155.0 MILLION TO THE REPUBLIC OF TURKEY FOR AN INDUSTRIAL TECHNOLOGY PROJECT

December 19, 2006 Private and Financial Sector Department and Turkey Country Unit Europe and Central Asia Region Public Disclosure page 11 of document ; page 15 of pdf

[51] http://web.worldbank.org/external/projects/main?pagePK=64283627&piPK=73230&theSit ePK=40941&menuPK=228424&Projectid=P080827

Program ID P080827 US$505.05 MILLION SERIES OF PROGRAMMATIC LOANS FOR SUSTAINABLE AND EQUITABLE GROWTH LOAN I

The World Bank to the FEDERATIVE REPUBLIC OF BRAZIL

F. Results Framework Analysis Program Development Objectives (from Program Document) pg v

Brazil First Programmatic Loan for Sustainable and Equitable Growth—P080827

	Original	Actual
Sector Code (as % of total World Bank financing) [52]		
Banking	25	25
Central government administration	10	10
General industry and trade sector	35	35
General transportation sector	25	25
Law and justice	5	5

According to the World Bank data, some of the achievements were as follows:

Indicator 1.   "Custom selectivity levels of 40% cut from 40% to 30%, and average gross release time for merchandise cut by 20%"

Indicator 2.   "Actions to increase efficiency of docks taken Port Productivity Improvement Plan approved"

Indicator 3.   "Output based maintenance contracts on 15% of federal road networks" improved to "output based maintenance/rehab contracts on 30% of federal road networks"

Indicator 4.   "Geographical restructuring of railways underway" Improvement observed was a "10% increase in non-road transport share." "Geographical restructuring of railway concessions completed"

Indicator 5.   "PPP Law submitted to Congress. PPP Law approved by Congress and Career Development Plan for Regulators submitted to Congress" (PPP, Public-Private Partnership)

Indicator 6.   "Amendment to Antitrust Law reviewed by inter-ministerial committee Amendment not sent to Congress." However, it was noted "some improvements made in enhancing competitiveness environment. Premerger notification to CADE has been made mandatory and CADE has investigated and decided on several mergers and acquisitions."

Indicator 7.   "Constitutional amendment approved to unify tax collection at federal, state and municipal levels for micro and SMEs." "Law regulating the unification of tax collections approved by Congress," "this law, approved by Congress in December 2006, unifies tax regime for micro and SMEs. Tax exemption for capital goods imports of exporting firms also approved."

Indicator 14. "Evaluation completed of operations and management procedures of Sector Funds and FINEP." "Regulation of Fondo Verde Amarelo and other mechanisms to support R&D introduced, also tax breaks and tax incentives for private R & D developed."[52]

G20 and World Bank Financing of Space Development

Now we need to look into a possible method by which the G20 community and the World Bank will provide a good portion of the funding for the Development of Space. Just as 5 solar systems were considered in the space credit determination of the future value of gold to be acquired. We must look at these same 5 solar systems and the future value of their GDP in order to come up with a sum the geographic economic communities will be able to finance against to obtain the space development funding. In this calculation we will need to assume that each future solar system will eventually have an equivalent or better economy than we currently have on the earth. The IMF has indicated that on a Global level the GDP of the earth's economy is annually close to $60 Trillion U.S.

---

[52] http://web.worldbank.org/external/projects/main?pagePK=64283627&piPK=73230&theSit ePK=40941&menuPK=228424&Projectid=P080827

Program ID P080827 US$505.05 MILLION SERIES OF PROGRAMMATIC LOANS FOR SUSTAINABLE AND EQUITABLE GROWTH LOAN I

The World Bank to the FEDERATIVE REPUBLIC OF BRAZIL

F. Results Framework Analysis Program Development Objectives (from Program Document) pgs iii, v, vi, vii, and viii

World Gross Domestic Product, current prices U.S. dollars Trillions

Year	2008	2009	2010	2011	2012	2013	2014	2015
World GDP U.S. dollars Trillions	61.2	57.9	61.7	65.0	68.7	72.7	77.1	81.7

Source:
IMF International Monetary Fund (and Forecast)
World Economic Outlook Database, April 2010 [53]

In order not to be over optimistic we should conservatively estimate each future solar systems approximate GDP at $50 Trillion U.S. The total GDP for the 5 solar systems together would be $250 Trillion U.S. Therefore the G20 community and the World Bank would therefore need to set up some mechanism whereby each geographic economic community could access financing against 20% of that total GDP of $250 Trillion. The total available to finance against would therefore be $50 Trillion U.S. This financing would have to be at a very minimal rate and be over a very long period of time. Since the timeframe of funding the startup of space development till the point at which active solar system economies are up and running will likely be in the hundreds of years at least.

The question may arise why we should put this debt burden upon the new solar systems. This is not only necessary, but fair, in order for the space development to occur. A lot of the populace that will colonize the new solar systems will be from the earth. Many of whom will likely be disadvantaged from an employment opportunity perspective. As well the means of setting up habitation on these new planets will be supported by the earths funding, technologies, and resources. The earth's current population will be essentially funding the development of the future solar system colonies. Therefore it is only right that the earth's populace

---

[53]   http://www.imf.org/external/pubs/ft/weo/2010/01/weodata/weorept.aspx?sy=2008&ey=2
015&scsm=1&ssd=1&sort=country&ds=.&br=1&c=001&s=NGDPD&grp=1&a=1&pr.
x=51&pr.y=9 IMF International Monetary Fund World Economic Outlook Database, April
2010 4. Report for Selected Country Groups and Subjects

should get the benefit of eventual revenues from investing a great deal of funding for the development of these 5 future solar systems.

Another factor to consider is how much of the funding will be directly funded and financed by each geographic economic community. In most instances of the current World Bank program financing, the nation involved puts up the greater portion. While the World Bank provides a portion of the financing as well as a great deal of administrative and procedural expertise. How the financing or funding will actually happen is yet to be determined.

Space Development Industry

Given that the World Bank, in coordination with the G20 economies, proceeds with either (a) financing the space development and or (b) funding space development with the space credit currency, then the next step to follow would be for each economic community to set out their respective economic plan of implementation.

I would assume that the World Bank would spread out funding (the USD$4 trillion) space credits over time and start with a smaller initial annual amount. This, of course, would need to be distributed internationally to each geographic economic community. Those without a formal economic community structure would need to make arrangements (work out a program with the representatives of all the nations from that geographic economic community).

The economic plans for development should encompass the following areas: (a) expanding the various national or economic community space agencies programs in coordination with the supporting industries, (b) rationalizing the cities and industry infrastructure, and (c) environmental cleanup. The respective plans, once approved by the World Bank, would then be funded for implementation in each geographic economic community.

Space Development Agencies

Forgive me for not including the smaller space development companies or agencies as well. For the purpose of this exercise, I will only be considering the "official" space agencies that effectively support the various geographic economic communities even though there are contributing national agencies involved as well.

The tremendous work these agencies are doing currently, with the limited funding they are getting, is a testament to human ingenuity even though the limited funding being provided today is no small amount considering the budgetary restraints many of the contributing nations are currently dealing with.

Geographic Economic Community Space Agencies: Reported Budgets

NASA current annual budget approx   USD$18 billion (North America)[54]

ESA 2009 Budget (approx. EUR$3.6 billion)   USD$5 billion (Europe)[55]

Russian Space Agency (approx. 58.2 billion RUB)   USD$2.2 billion (Russian Federation and associate states)[56]

Chinese Space Agency approx.   USD$0.5 billion (China)[57]

The current main activities associated with space development and the space agencies include the following: technologies, launchers, exploration, human spaceflight, navigation, telecommunications, Earth observation, science, and more.

---

[54] http://www.nasa.gov/news/budget/index.html
FY 2010 Budget Estimate (9.12 MB PDF
3452225main_FY_NASA Budget 2010_UPDATED_final_5-11-09_with_cover
National Aeronautics and Space Administration President's FY 2010 Budget Request Detail pg ix

[55] http://www.esa.int/SPECIALS/About_ESA/SEMNQ4FVL2F_0.html
Funding : ESA budget by programme for 2009

[56] http://en.wikipedia.org/wiki/Russian_Federal_Space_Agency

[57] http://en.wikipedia.org/wiki/Chinese_Space_Agency

Space Development, Our Future

All the above main activities would need considerably more funding to speed up the process of space development in order to get to the point of viable spaceflight and space development. Those geographic economic communities, with limited presence in the space industry, would of course need funding to bring them eventually close to par with the developed economic communities. A very important consideration in all the funding being provided is that each geographic economic region has its own scientific and philosophical contribution to make. Without ensuring that all geographies and communities are participating, we would most likely be missing out on certain perspectives that would diminish the extent and scope of the scientific and economic developments.

The main areas we would initially need to focus on are developing and setting up advanced space stations around Earth, Mars, Venus, and any of the larger planets' moons and expand our satellite and communication capabilities near and around these planets as well. Not only that, but we should put considerable investment R&D into other possible methods of space transportation other than the conventional rockets (if that is possible).

While we still have a viable world economy, we need to realize the implications should we allow the world economy to decline or stay in an at-risk situation for any extended period of time, the most serious implications being potential losses in skilled workforce, infrastructure, R&D, and high-tech industries. A good example of this is as stated in the EU Commission statement from the Economic and Financial Affairs Directorate, while directed at the EU states it is applicable to the other developed and developing nations in the world.

> The crisis may accelerate downward pressures on trend growth. The Commission had projected that potential GDP growth in the euro area would fall in the long run due to the ageing

population. A number of crisis-related factors may amplify this phenomenon. First, unemployment, if protracted, would entail a prolonged, perhaps permanent, loss of valuable skills. Second, the stock of equipment and infrastructure will decrease and may become obsolete due to lower investment and sectoral change. Third, innovation may well be hampered as research and development spending are usually among the first outlays that businesses cut during a recession. Higher risk premia may make the financing of R&D more expensive in the future. The loss in potential growth is expected to be higher in countries experiencing deep recessions.

The reduction of divergences within the euro area in the immediate aftermath of the crisis is welcome. In the immediate wake of the financial crisis, growth took a dive in all euro area countries, but to differentiated extent. The Commission services interim forecast shows that growth trajectories are beginning to fan out within the euro area. For example, growth in 2009 has been revised upwards for Germany and France, while Italy and Spain recorded downward revisions. As for divergences in current accounts, the ongoing housing market correction and its impact on domestic demand is likely to go some way towards reducing disparities, a welcome step towards more balanced growth patterns. However, the convergence is moderate and is not consistent among the euro area Member States.[58]

Another main consideration in the funding to be provided to the geographic economic community space agencies is the cumulative and long-term benefit the space industry has on the overall economy, in employment for skilled workforces, a market for contributing industries, a driver for R&D, which brings about increasing technological improvements. A good example of this is stated in the Oxford paper "The Case for Space:

---

[58]   http://ec.europa.eu/economy_finance/publications/publication15951_en.pdf
European Commission Annual Statement on the Euro Area—2009 Economic and Financial Affairs Directorate (pg 4 in report)

The Impact of Space Derived Services and Data." As a case in point for the UK space industry.

> "The technological advances that come about as a result of R&D investment in the space industry can be transferred to firms in other sectors in the form of 'spillover' effects. Previous research by Oxford Economics suggests that such spillover effects are very large, with R&D investment by the aerospace sector generating a social return of around 70%—i.e. every £100 million invested in R&D leads to an increase in GDP of £70 million in the longer term. On this basis, we estimate that the space industry helps to generate £900 million a year of GDP in the UK due to the spillover effects of its R&D, on top of its £5.6 billion of direct and multiplier impacts. So, we calculate that the space industry overall currently contributes at least £6.5 billion a year to UK GDP."[59]

NASA's 1960 moon missions, in that time frame, were a concerted effort of the U.S. government to race to the moon to supersede efforts by the Soviets. While initially politically and cold war motivated, the funding at that time was considerable. Later on, the NASA and space development efforts were diluted due to more pressing tasks such as funding the United States' efforts in the Vietnam War.

Currently, the various space community agency space development efforts are limited to the creation and expansion of the international space station as well as some initial space-scientific missions, such as the Mars lander and various satellite missions, which is no small commitment but probably not enough to generate a real growth strategy for the new economic area of space development.

Should the G20 community financing for space development proceed, we will not only have a moon project to develop but various projects for Mars, Venus, and some of the asteroids, in addition, further scientific missions, developing other methods of transportation, and hopefully eventually missions out of our solar system. This will likely be of great benefit

---

[59]   The Case for Space: The Impact of Space Derived Services and Data Final Report—July 2009 Commissioned by South East England Development Agency (SEEDA) Oxford Economics 121, St Aldates, Oxford, OX1 1HB ?: 01865 268900, : 01865 268906: www.oxfordeconomics.com case_for_space UK Oxford.pdf Section 1 Executive Summary   pg4 of pdf

to related industries, supporting industries, and employment numbers all around. The World Bank G20 funding needed will be great, but we must keep in mind the future payoff will be even greater once we start generating revenues and accumulating the wealth from space development.

City and Industrial Infrastructure

Now we come to deciding what areas in cities and industries need restructuring. For cities, the main focus should be on continuing the trend for high-rise accommodation and office space, recovering the older, outdated housing/buildings and, as needed, totally modernizing the waste treatment processes and facilities. In addition, for any secondary education facilities that would contribute to the space industry, where needed, to provide funding for any needed updating and resources. For industries, they would be provided funding for additional resources required. As well, should there be improvements that can be funded to improve the functionality of the industrial sites so as to significantly reduce the pollution in their production process, then this should be done as well.

Cities

Cities will need help in order to support the greater populations that now reside in the large metropolitan areas. It will require a great deal of cooperation at all levels to achieve much greater efficiencies in urban design to prevent further urban sprawl which is proving very costly to maintain and very harmful to nature and wildlife. Many urban areas that have deteriorated or are quite old need to be revitalized into high-rises or redeemed to nature. The waste disposal systems need to be redesigned in terms of efficiencies and prevention of pollution. The overall municipal design, urban transportation systems, and total infrastructure need to be as efficient and as reasonable in cost as possible to maintain. The infrastructures also need to be updated as per schedules not when the update is past due and environmental damage occurs. In addition, while nuclear power may not appear to be the best choice environmentally, the newer designs, I can only assume, are far better overall than keeping older designed units in production. The choice would be clear to implement new units as soon as can be done and remove the older units in as timely a fashion. The rationale for this is that going forward we will need to focus our resources on space development rather than continually putting resources into inefficiently designed cities. Cities that will constantly need attention paid to their infrastructures will result in an unnecessary drain on much-needed financial resources.

The Need for the Evolution to High-rise Accommodations versus Single Homes

Given that we humans share our planet with animal and plant life and that they require much more living space than we need to survive, we as the species that have dominion over, and responsibility for, the animal and plant life on this world need to change our lifestyles to be more accommodating to their needs. This will require us to evolve toward cities that are more centered toward high-rise living space, high-rise office and retail space, and especially a more efficient industrial land use.

To highlight this point, we will examine the most prevalent urban land use, which are single-family homes. We will compare this land use to that taken up by high-rise apartment use.

A. Generic North American Single Home:
    Lot Size 50 ft. × 100 ft. or 5,000 square feet (to simplify the calculation)
    House size 2,500 square feet House dimensions 30 ft. × 42 ft. approx.

B. Generic North American Apartment Building:
    Lot Size 66 ft. × 660 ft or 43,560 square feet (equal to 1 acre)
    Apartment Size 2,200 square feet Apartment dimensions 30 ft. × 73 ft. approx.

    In this example, we will assume a fourteen-story dwelling with approximate dimensions of 75 ft. × 146 ft., with four family apartments per floor. This provides a total of fifty-six apartments in this building.

C. Comparison of a Single Home vs. Apartment Building Land use:
    The land space used by the apartment building would accommodate the space required for approximately nine single homes whereas the land space used by the apartment building is able to accommodate a total of fifty-six family apartments (in this example). So the land is used six times more effectively with apartment buildings being used rather than single homes.

To expand on this further (continuing with the above examples), if four apartment buildings were built and used rather than the equivalent two hundred single homes, the land space used would be 1,000,000 square feet for the two hundred single homes versus 174,600 square feet for the four apartment buildings. We would be preserving wildlife land and or agricultural land of about 825,000 square feet or about 18 acres.

I have personally seen deer, foxes, and geese trying to return to land that was once theirs, land that has been developed for single houses. It is not a good feeling seeing them never being able to return to what was once theirs. We can make a difference in our lifestyle choices. If we make the right choices we can at least put a stop to the loss of nature to housing or industrial development.

Industries

Industries mainly benefiting from space development would be high-tech industries requiring well-trained and or highly qualified employees. While many employees from the automotive industry are unemployed, opportunities should be provided to these and other skilled workforces to move into the space industry field and, if needed, training provided to meet the skill sets required.

I expect many industries already in the space development field (or entering it) such as satellite manufacturers, robotics industry, those currently involved in the space station, and those currently involved in scientific space missions will find their funding and support to increase drastically should the space credit and long-term financing for space development be initiated.

In many parts of the world where there are industries and factories in place that are deemed big polluters, arrangements should be made to modernize these facilities and processes to make them efficient and nonpolluting. This would likely be done as part of the fiscal arrangements between the national or economic community and the World Bank and the G20.

Environmental Cleanup and Prevention

The main focus areas for pollution cleanup should be municipal waste dumps, industrial production and mining sites, and ocean dump sites (including those locations waste has drifted to). All these areas for pollution cleanup need to be funded with a portion of the space development financing funding and or the subsequent increased national tax base.

With municipal waste sites, the waste can be separated into compost, plastics, metals, other, and hazardous wastes. Each waste type can be dealt with in the appropriate manner of recycling or destruction. Industrial production and mining sites that need to be cleaned up should be provided the necessary funding for the appropriate cleanup (whether hazardous metals or chemicals). Ocean dump sites or sites where the pollution has drifted to need to have the waste recovered and recycled as appropriate to the type of waste(s).

In terms of prevention, the municipal/industrial production and consumption process needs to be extensively revamped in terms of increasing the extent of recycling that can be done. While in many cases, products being produced today are more readily recyclable. We still have a long way to go.

The large municipal waste sites, rather than just having the waste dumped, should have the waste processed into each type of product and the respective appropriate environmentally friendly recycling or processing.

The existing waste sites on land and sea need to be revisited to see what can be reprocessed in an environmentally friendly manner. While cleanup and updating to environmentally friendly processing methods will likely be a costly process, we will be further developing existing waste processing industries that can also contribute to the economies of developed nations. In addition, newer waste processing industries may be established.

Rationale for Immediate Action

Other than the development of space providing the world's economies with the needed stimulus to have a truly prosperous economy, we also have to look at other reasons why we need to develop space and soon. Earth has experienced many catastrophes in the recent past from asteroid collisions, climatic changes, and polar shifts, to ice ages. The question we must face is, where are we in the timeline for the next occurrence? Not only that, but when would we be ready in the event of a truly catastrophic occurrence to evacuate a good portion of the populace and be in a position to terra-form another planet and inhabit that planet?

Given that we use the hidden wealth (space credits) and financing to develop space and soon, with this tremendous influx of financing, we should be able develop our technology to a much higher level. Hopefully in sync with that technological development, we should also encourage development in our spirituality, philosophy, experiencing the arts, knowledge, and understanding. After all for our civilization to really grow to its potential we must strive for a better quality of life for more people.

We should take into consideration what happened to the native populations of the Americas and compare it to the prevalent attitude that we are alone in the universe. Essentially, in the past, Europeans came out of nowhere in large numbers to the Americas to establish colonies and develop the resources. Is it possible our human civilization might face the same fate in the not-too-distant future? With a more developed technology on Earth, we would most likely be in a better position to face such a challenge in the event a similar unexpected arrival should occur on Earth from elsewhere.

While we believe the ice age occurred fifteen to twenty thousand years ago, there is still a lot of uncertainty about our history beyond four to five thousand years ago. A lot of ancient cultures report a massive flood that appears to have been on a worldwide scale. When this actually occurred has still not been determined. Some of these ancient cultures report also prior civilizations and catastrophes to the flood. In any event, the main consideration here is that there is a lot of historical information that is not known (as to what actually occurred and when). There are also a lot of variables and dynamics at play with our planet Earth—the sun and the possibility of a minute change in strength (magnetic or warmth) of the sun that could have a significant effect on Earth, our Earth's own magnetic field and possibilities of polar shift(s), Earth's thin crust and the potential for crustal displacement, climate change, elliptical change in orbit around the sun—many things must be considered. It is known that the polar arctic zones of Earth were at one time the tropical zone. Most likely, a change occurred in the magnetic field and a polar shift occurred. In reality, most of us are living in blissful ignorance of what to expect while we hurtle through space.

NASA is charged with seeking out nearly all the asteroids that threaten Earth. The U.S. Congress assigned the space agency this mission. To date, NASA has not been provided much of the money to build the necessary telescopes. NASA was requested to spot 90 percent of the potentially deadly rocks hurtling through space by 2020. "NASA says it has completed about one-third of its assignment with its current telescope system. NASA estimates that there are about 20,000 asteroids and comets in our solar system that are potential threats to Earth. They are larger than 460 feet (140 metres) in diameter—slightly smaller than a sports stadium in New Orleans. So far, scientists know where about 6,000 of these objects are. Rocks between 460 feet and 3,280 feet (1,000 metres) in diameter can devastate an entire region but not the entire globe, said Lindley Johnson, NASA's manager of the near-Earth objects program. Objects bigger than that are even more threatening, of course." Recently, "astronomers were surprised when an object of unknown size and origin bashed into Jupiter and created an Earth-sized bruise that is still spreading. Jupiter does get slammed more often than Earth because of its immense gravity, enormous size and location."[60]

That is truly an important consideration, the recent Jupiter collision from an object of "unknown" origin. Hence, the need for space stations and a multitude of satellites well beyond Earth. Our technology is also not yet in a position to deflect the trajectory of any such asteroids in the event of a potential collision with Earth. We should complete the funding for the work needed to make that possible.

The population of Earth as of 2010 is estimated to be about 6.8 billion (mid-2009). This population and continued population growth presents challenges. The most pressing challenge is feeding our world population a viable amount of food. But we also face increasing problems with pollution, employment, social security support, habitable land, and remaining energy sources. Hopefully, in the short term, at least we can educate the populace (young adults) in family planning as well as educate about contraception (or alternatives for those whose religions prefer that route) until such time as space transportation is viable to habitable (hopefully unoccupied) planets in our galaxy.

Recent World Population Figures[61]
1900 1.6 billion
1950 2.55 billion
1990 5.3 billion
1999 6 billion
2009 6.8 billion

Projected World Population Figures[61]
2011 7 billion
2025 8 billion
2050 9.4 billion

---

[60] http://ca.news.yahoo.com/s/capress/090812/science/science_us_sci_killer_asteroids
NASA's near-Earth object site: http://www.jpl.nasa.gov/asteroidwatch
[61] http://geography.about.com/od/obtainpopulationdata/a/worldpopulation.htm
2010 About.com, a part of The New York Times Company.
Current World Population and World Population Growth Since the Year One
By Matt Rosenberg, About.com Guide

It is estimated that the sustainable population Earth can support is approximately 5 billion people according to Optimum Population Trust (a green think tank). This 5 billion number is assuming there would still be a sizeable number of people living below the poverty line. The sustainable criteria used is "an optimum population means, at its simplest, a population size which is environmentally sustainable in the long term, affords people a good quality of life, has adequate renewable and non-renewable resources necessary for its long-term survival and consumes or recycles them to ensure it will not compromise the long-term survival of its progeny."[62]

So until such time as we are able to develop space travel and development to other planets for large-scale human habitation, much of the population on Earth will be under severe conditions of constraint (food, habitation, and material wealth). Much of the population of Earth is already under those constraints, and further population growth will only add to this negative situation. It would appear humanity is blessed by our capacity to reproduce while at the same time cursed with that capacity while we are constrained by the limitations of only living on this Earth and not being able yet to travel to other habitable planets.

While any ice ages, climatic changes, or polar shifts may appear to be way off in the foreseeable future, we are still, in space-development terms, nowhere near ready to evacuate the populace and "nature," transport the same, or be able to go to a habitable planet in transportable proximity. How long it will take us to be in a position to undertake such a task, we don't know. It could be decades, most likely centuries or, even possibly, millennia. While our technology is significantly more advanced than the sailing seafaring days, in the hostile environment of space where the distances involved are so great, our technology to date is relatively quite limited. To develop the technology and knowledge required will take, I assume, a considerable length of time. So the sooner we start a committed and concerted effort in space development, the less our civilization on Earth is at risk.

---

[62]   http://www.optimumpopulation.org/opt.optimum.html
Optimum Population Trust Towards sustainable and optimum populations
Section Defining an optimum population

Conclusion

While many may feel space development is a luxury or should be reserved for science fiction novels and the movies, our human destiny lies in the direction of space development. Based on many civilizations' historical development, we have always expanded geographically in an expansionary manner. Whether we take the necessary steps now or wait a hundred years or more, it would appear inevitable that we humans will expand into space.

From a historical perspective, we on Earth have reached similar conditions to those that were prevalent in Europe in the 1300-1400s. At that time in Europe, one of the main conditions was overpopulation mostly in the rural areas but even in the cities. There was a rise in poverty and unemployment with little hope for those in dire straits. Fortunately, in our modern economy, we have social programs that help those in need. But these social programs are being strained, with not all being able to take advantage of them, and with the conditions not likely to improve. There are also many locations on Earth where there is a great deal of poverty with little opportunities for gainful employment. We will need to educate our young on the serious necessity of family planning (worldwide) given Earth's current population and continued growth beyond Earth's capability to support the population sustainably. But it is also apparent we have reached that stage in history, whether it is now or in the near future, where we need to colonize space. Just as in Europe of the 1400s, we will need to implement programs to explore and develop space. We will be setting the stage for habitation and colonies on other planets as well as obtaining the wealth needed for growing the world's economy.

Space development, as I have indicated above, is an urgent necessity not only from Earth's population standpoint, but also from the standpoint of creating the ability to provide habitation on other planets, which will give our human civilization options should we find Earth at risk from the many and varied events that could happen. The risk being based upon both historical events and scientific projections using valid data.

If the various economic communities and nations make a serious effort and produce well laid-out plans for development, the space industry will be a dynamic addition to the world economy as well as offering unlimited growth potential in time. The financing options laid out in this book should enable the world economic communities and or nations to be able to finance space development without placing serious debt burdens on the participating nations. Why let an opportunity like this pass by when we can create and generate a market for many industries that are seeing their markets currently at a virtual standstill or at risk? Most important of all, we need to consider the need for gainful employment for all workers in the many sectors of industry and regions of the world. As well, the space development industry should not only offer opportunities to developed nations, but to developing nations, whether directly or indirectly.

Index

A

asteroids, 8, 58
  belt, 40–41
  collisions, 9, 62

B

Brazil First Programmatic Loan, 50
BSE SENSEX index. *See* stock index

C

"Case for Space, The" (Oxford
    Economics), 57
conditions, urban and industrial, 27
Crisis, Finance, and Growth 2010
    (World Bank), 20, 47

D

debt, 22, 26, 34
  constraints, 14
  national and international, 27

E

Earth
  atmosphere on, 36
  catastrophes on, 8
  cleanup of, 38
  data of, 42
  resources on, 7
  wealth on, 40–41
economic activity, 16
economic conditions, 28, 29–30,
    31–32, 33
  current, 11, 19, 34
  preventing erratic, 40
Economic Cooperation Act. *See*
    Marshall Plan
economic downturn, effects of, 20
economic growth
  of Brazil, 49
  of China, 31
  a new avenue of, 8
economic plans, 54
economic prosperity, 14, 40
economy, 15

active, 43
of Brazil, 51
dynamic addition to world, 68
of Europe, 13, 15
of India, 25–26, 31
of Japan, 24
of Russia, 26, 33
space, 8
turmoil in the world, 11
U.S., 20
employment
increase in, 49
loss of, 28
environment
business, 49
cleanup of, 34, 44, 54
conditions of the, 34–39
economic, 28
problems for the, 38–39
space, 7
Euro area, quarterly report on the, 13
European Commission, 13–14, 56
European Economic Recovery Plan
    (EERP), 15
Excessive Deficit Procedures (EDPs),
    15

F

financial community, 7, 44
financial crisis, 16, 20, 56
financing, long-term, 7, 40
FTSE. *See* stock index
funding
  G20, 58
  infrastructure, 28
  space credits, 27, 40, 43
  space development, 40

G

G20, 43–44
  financing, 7, 34, 44
  fiscal stimulus plans, 27
  space credits and, 43
  space development and, 54, 58
global activity, 16, 26
gold, 7–8, 40–41, 43
government
  Chinese, 31
  European, 7
  Japanese, 32
  Russian, 26
  spending, 25
  U.S., 57

H

Hang Seng index. *See* stock index
high-rise accommodations, 58–59

I

IBRD (International Bank
    for Reconstruction and
    Development), 46
IFC (International Finance
    Corporation), 30
IMF (International Monetary Fund),
    24, 26
World Economic Outlook, 26
industry restructuring, 34, 44
infrastructure
  cities, 58
  industry, 54
  losses in, 56
  Mumbai's challenge in, 30

space development, 41, 43–44
Tokyo projects, 32
World Bank projects, 45
infrastructure funding, 7, 28, 34

J

job losses, 21-22. *See also*
    unemployment

K

Kommersant, 33

L

lost generation, 22

M

market conditions, 19
Marshall Plan, 44-45

N

NASA (National Aeronautics and
    Space Administration)
  moon missions, 57
  solar system data, 42
Nikkei index. *See* stock index
North American Free Trade
    Agreement (NAFTA), 11

O

Oxford Economics
  "Case for Space, The," 57

P

pollutants, 8, 37
pollution, 8
  air, 36
  cleanup, 8, 61
  forms of, 34, 38
  land, 37
  ocean, 34, 36
  prevention and solution, 34, 38-39,
      44, 58
  problems with, 63
population, 23, 26, 28-33, 47, 55, 63-65
  conditions of constraints, 65
  growth, 64-65
  world, 64

R

recession, 15, 20, 56
  effects of, 30
  effects on Indian economy, 31

S

solar systems, 7–9, 40–43, 58
space, expansion into, 7, 43, 67
space agencies, 55, 57–58
space credits, 27, 43–44
  basis of, 7
  funding with, 40
  World Bank and, 54
space development, 58–59, 67–68
  areas to focus in, 55-56
  funding for, 40, 43, 54
  growth potential of, 27
  infrastructure and, 44
  our future, 55–58

space development funding, 34
space development industry, 54, 68
space industry, 8, 27, 57
  addition to the economy, 68
  viable, 40
space stations, 56
stimulus
  fiscal, 15, 24–27
  monetary, 24
stock index, 11–19

T

terra-forming, 41

U

unemployment, 28, 56
  decrease in, 27
  rate of, 21–23

V

Venus
  data of, 42
  terra-forming, 41
  wealth in, 40–41

W

waste, 34
  four main types of, 35
  radioactive, 35
  storage of, 38
Waste Management Corporation,
    38–39
wealth
  current world gold, 40
  hidden, 41, 43
  off-world, 8, 41
  space development, 58
World Bank
  Crisis, Finance, and Growth 2010, 20
  economic forecast, 24
  financial community, 7
  financing, 8, 27, 34, 40
  projects, 45–47, 49–51
  space credits and, 43
  space development and, 54
World Economic Outlook (IMF), 26
world economy, 11
  decline of, 56
  setup, 34
  space industry and, 40, 68
  underutilizing, 27